MORMONISM
AND EVOLUTION

MORMONISM AND EVOLUTION:

THE AUTHORITATIVE LDS STATEMENTS

William E. Evenson
&
Duane E. Jeffery

GREG KOFFORD BOOKS
SALT LAKE CITY   2005

Published by Greg Kofford Books, Inc.
Salt Lake City, Utah.

www.koffordbooks.com

2009 08 07 06   5 4 3 2

Pictures courtesy of Church Archives. The Church of Jesus Christ of Latter-day Saints.

Library of Congress Cataloging-in-Publication Data

Mormonism and evolution : the authoritative statements / Duane E. Jeffery & William E. Evenson.
        p. cm.
  Includes bibliographical references and index.
  ISBN-13: 978-1-58958-093-0 (alk. paper)
  ISBN-10: 1-58958-093-1 (alk. paper)
  ISBN-13: 978-1-58958-097-8 (limited leather edition)
  ISBN-10: 1-58958-097-4 (limited leather edition)
  1. Evolution (Biology)--Religious aspects--Church of Jesus Christ of Latter-day Saints. 2. Church of Jesus Christ of Latter-day Saints--Doctrines. I. Evenson, William E., 1941- II. Title.
BX8643.E94J44 2005
231.7'652088289332--dc22
                        2005028855

Table of Contents

Illustrations

Front Cover

1. Gregor Mendel
2. James E. Talmage
3. Charles Darwin
4. Heber J. Grant
5. Charles Lyell
6. Spencer W. Kimball
7. David O. McKay
8. Joseph Fielding Smith
9. John Scopes

Introduction

The *BYU Evolution Packet* and Cover Sheet

William E. Evenson

The *BYU Evolution Packet*, "Evolution and the Origin of Man," was assembled and made widely available on the BYU campus in 1992 as a result of the large volume of student questions on this subject. As at other universities, the subjects of evolution and the origin of man are treated at BYU in courses in several different departments. Given BYU's LDS Church sponsorship, students in these courses naturally wished to know the official position of the Church of Jesus Christ of Latter-day Saints on this subject. Consequently, some faculty members both in the sciences and in religion had gathered material on these topics to distribute to their students, with the natural result that students might receive one set of statements by Church leaders from one professor and a different set from another professor.

Several faculty members and administrators

felt that the diversity of materials on these subjects, which were often selected to emphasize the views of the professor, tended to create confusion in the minds of students and accentuate the potential for controversy about the Church's position. While there have been disagreements about the correct view of evolution and the origin of man among individuals, there has been remarkable good will on all sides among faculty and administrators.

In 1991, in response to questions from students about the Church's position on evolution, BYU President Rex E. Lee requested that a packet of authoritative information be placed in the University Reserve Library as a source to which students could turn for information about the Church's position on evolution and the origin of man. This set of materials was assembled by biology professor Duane E. Jeffery.

Because of my experience in preparing the article on "Evolution" for the *Encyclopedia of Mormonism,* and in my capacity at that time as Dean of the College of Physical and Mathematical Sciences, University Provost Bruce C. Hafen asked me to consider a packet that could be made available to students as the "official Church position" on this subject. As the cover memo provided by the BYU Board of Trustees clarifies, "formal statements by the First Presidency are the definitive source of official Church positions." Therefore, these formal statements would provide the foundation for the packet that was assembled. I also considered whether other, quasi-official statements should be included, as explained below.

It was immediately clear to me that the selection of material for such a packet could not depend on the content of the statements that were included. It must depend only on the official status of the statements themselves; i.e., all statements that can be said to be "official" should be included and others excluded, irrespective of how much that collection might support or undermine any particular point of view. The goal was not to achieve some kind of "balance" among the views that have been expressed, but to give students of this subject the full range of official views so that they could evaluate for themselves the different positions they might encounter. These official views therefore provide the basis for evaluating other expressions that are not themselves authoritative.

In accord with this philosophical stance, I prepared an initial draft of the packet which contained all formal First Presidency statements relating to this topic. I also collected all published statements made by the Presidents of the Church during the time that they held that office. I considered the speech, "The Earth and Man," given in 1931 by Elder James E. Talmage of the Quorum of the Twelve, because that speech was reviewed and approved by the First Presidency and officially published by the Church. Finally, this draft packet included the *Encyclopedia of Mormonism* article because of the excerpt from First Presidency minutes in 1931 addressing the Church's stance towards scientific studies of evolution and the origin of man. The packet thus was made up entirely of materials having some level of authoritative status, i.e., First Presidency, President of the Church, or published officially by the Church.

The contents of this draft packet were reviewed by Dean Robert L. Millet of Religious Education, Dean

Clayton S. Huber of Biological and Agricultural Sciences, and Provost Hafen. After considerable discussion Provost Hafen sought counsel from members of BYU's Board of Trustees. They took the position stated explicitly in the cover memo to the packet, namely, that the University packet should contain only those items that represent the official position of the Church, i.e., formal statements by the First Presidency as a body. The *Encyclopedia* article was left in because of the First Presidency minutes included in it, which are not otherwise available to the public. This final packet, with a proposed cover sheet, was then reviewed by BYU's Board of Trustees consisting of the First Presidency, seven members of the Quorum of the Twelve, and other General Authorities and Officers. In June 1992, they approved the packet for distribution to the university community and provided the final version of the cover statement to introduce the packet to its readers.

I emphasize that balance was not the issue in preparing the packet. The issue was providing only those materials that could clearly be said to impart the official, declared position of the Church. The cover statement was drafted initially by me with subsequent revisions by the BYU committee consisting of Provost Hafen and Deans Millet and Huber as well as myself. The draft we provided was used as a basis for the final version approved by the Board, but they reworked it substantially, so that the final paragraph and the clarification that "formal statements by the First Presidency are the definitive source of official Church positions" came directly from that authoritative body.

How is the packet to be used at BYU? There

was never any intention in preparing this packet for Board review to place restrictions on professors who might wish to distribute other materials for their students. The Board and university administrators only requested that BYU faculty members refer students to the materials in this specific packet whenever relevant subjects were addressed in classes and that this packet also be distributed *along with* any other items they may choose to distribute. Furthermore, when other items are distributed, they should be clearly separated and given as a supplement to this material and include a fair sampling of the diverse viewpoints among LDS Church leaders. For example, if one included statements by LDS apostles in a handout on evolution, the range of views would include some statements against evolution, some sympathetic to evolution, and several shades of opinion in between. The Board agreed that BYU faculty should avoid the implication that a greater sense of unanimity or resolution of this topic exists than is actually the case.

Preface

In 1999 the four basic documents of the Packet **(1-4)** were distributed to all teachers in the Church's seminaries and institutes (the Church Education System). The *BYU Evolution Packet* itself is provided here for a wider audience beyond the University in the hope that other members of the Church will also have the foundation documents on evolution and the origin of man available as they study these topics. It is of crucial importance that members of the Church understand the primary role of "formal statements by the First Presidency" in establishing official doctrine. These are the statements that can be relied upon in the face of diverse unofficial views from Church leaders and others, views that may, in fact, be contradictory. Such contradictions, however, are no more than indications of different opinions on issues not fully settled by authoritative statements or modern revelation. Furthermore, all thoughtful, faithful Latter-day Saints share the responsibility to find their own insights on those questions that are not settled by formal statements of the First Presidency.

In the documents and commentary that follow, we reproduce first the entire BYU Packet; the individual documents are numbered 1,2,3,4. The Cover Page itself is self-explanatory, but we then include brief introductions to each of the four documents that make up the package proper. This is to provide historical context and continuity between the documents.

Following those documents we attach an "Appendix" which includes the other relevant documents pertaining to these issues and which were a) produced under the First Presidency's sponsorship, b) statements published by the President of the Church over his signature alone, or c) documents approved for publication directly by the First Presidency as a body. As with the Packet documents, these are given brief introductions by Duane E. Jeffery to provide historical and doctrinal context. They are given letter designations A through K.

BYU PACKET

Cover Letter of BYU Packet

October, 1992

EVOLUTION AND
THE ORIGIN OF MAN

This packet contains, as far as could be found, *all* statements issued by the First Presidency of the Church of Jesus Christ of Latter-day Saints on the subject of evolution and the origin of man, and a statement on the Church's attitude toward science. The earliest First Presidency statement, "The Origin of Man," was issued during the administration of President Joseph F. Smith in 1909. This was followed by a First Presidency Message in 1910 that included brief comments related to the study of these topics. The second statement, "Mormon View of Evolution," was issued during the administration of President Heber J. Grant in 1925. Although there has never been a formal declaration from the First Presidency addressing the general matter of *organic evolution* as a process for development of biological species, these documents make clear the official position of the Church regarding the *origin of man*.

This packet also contains the article on evolution from the *Encyclopedia of Mormonism,* published in 1992. The current First Presidency authorized inclusion of the excerpt from the First Presidency minutes of 1931 in the 1992 *Encyclopedia* article.

Various views have been expressed by other Church leaders on this subject over many decades; however, formal statements by the First Presidency are the definitive source of official Church positions. It is hoped that these materials will provide a firm foundation for individual study in a context of faith in the restored gospel.

Approved by the BYU Board of Trustees

June, 1992

Context

The year 1909 was a double anniversary: the 100th year of Charles Darwin's birth, and the 50th anniversary of the publication of his influential book *On the Origin of Species*. 1909 in particular brought a spate of statements and analyses pondering the validity and implications of his ideas. The LDS First Presidency had appointed a committee in 1908, chaired by Orson F. Whitney of the Quorum of the Twelve, to prepare a formal statement on the subject. That statement was published, over the signatures of the First Presidency, in November 1909, and constitutes the first of a series of statements signed or authorized by the First Presidency or President of the Church on the topic.

From left to right: President Joseph F. Smith, Second Counselor Anthon H. Lund, and First Counselor John R. Winder

(1)
First Presidency Statement
Improvement Era
13 (November 1909):75-81

Editor's Table.

The Origin of Man.

By The First Presidency of the Church.

"God created man in his own image."

Inquiries arise from time to time respecting the attitude of the Church of Jesus Christ of Latter-day Saints upon questions which, though not vital from a doctrinal standpoint, are closely connected with the fundamental principles of salvation. The latest inquiry of this kind that has reached us is in relation to the origin of man. It is believed that a statement of the position held by the Church upon this important subject will be timely and productive of good.

In presenting the statement that follows we are not conscious of putting forth anything essentially new; neither is it our desire so to do. Truth is what we wish to present, and truth--eternal truth--is fundamentally old. A restatement of the original attitude of the Church relative to this matter is all that will be attempted here. To tell the truth as God has revealed it, and commend it to the acceptance of those who need to conform their opinions thereto, is the sole purpose of this presentation.

"God created man in his own image, in the image of God created he him; male and female created he them." In these plain and pointed words the inspired author of the book of Genesis made known to the world the truth concerning the origin of the human family. Moses, the prophet-historian, "learned," as we are told, "in all the wisdom of the Egyptians," when making this important announcement, was not voicing a mere opinion, a theory derived from his researches into the occult lore of that ancient people. He was speaking as the mouthpiece of God, and his solemn declaration was for all time and for all people. No subsequent revelator of the truth has contradicted the great leader and law-giver of Israel. All who have since spoken by divine authority upon this theme have confirmed his simple and sublime proclamation. Nor could it be otherwise. Truth has but one source, and all revelations from heaven are harmonious with each other. The omnipotent Creator, the maker of heaven and earth—had shown unto Moses everything pertaining to this planet, including the facts relating to man's origin, and the authoritative pronouncement of that mighty prophet and seer to the house of Israel, and through Israel to the whole world, is couched in the simple clause:

"God created man in his own image" (Genesis 1:27; Pearl of Great Price-Book of Moses, 1:27-41).

The creation was two-fold—firstly spiritual, secondly temporal. This truth, also, Moses plainly taught—much more plainly than it has come down to us in the imperfect translations of the Bible that are now in use. Therein the fact of a spiritual creation, antedating the temporal creation, is strongly implied, but the proof of it is not so clear and conclusive as in other records held by the Latter-day Saints to be of equal authority with the Jewish scriptures. The partial obscurity of the latter upon the point in question is owing, no doubt, to the loss of those "plain and precious" parts of sacred writ, which, as the Book of Mormon informs us, have been taken away from the Bible during its passage down the centuries (I Nephi 13:24-29). Some of these missing parts the Prophet Joseph Smith undertook to restore when he revised those scriptures by the spirit of revelation, the result being that more complete account of the creation which is found in the Book of Moses, previously cited. Note the following passages:

And now, behold, I say unto you, that these are the generations of the heaven and the earth, when they were created in the day that I, the Lord God, made the heaven and the earth,

And every plant of the field before it was in the earth, and every herb of the field before it grew.

For I, the Lord God, created all things of which I have spoken, spiritually, before they were naturally upon the face of the earth. For I, the Lord God, had not caused it to rain upon the face of the earth.

And I, the Lord God, had created all the children of men, and not yet a man to till the ground; for in heaven created I them, and there was not yet flesh upon the earth, neither in the water, neither in the air.

But I, the Lord God, spake, and there went up a mist from the earth, and watered the whole face of the ground.

And I, the Lord God, formed man from the dust of the ground, and breathed into his nostrils the breath of life; and man became a living soul, the first flesh upon the earth, the first man also.

Nevertheless, all things were before created, but spiritually were they created and made, according to my word (Pearl of Great Price-- Book of Moses, 3:4-7. See also chapters 1 and 2, and compare with Genesis 1 and 2).

These two points being established, namely, the creation of man in the image of God, and the twofold character of the creation, let us now inquire: What was the form of man, in the spirit and in the body, as originally created? In a general way the answer is given in the words chosen as the text of this treatise. "God created man in his own image." It is more explicitly rendered in the Book of Mormon thus: "All men were created in the beginning after mine own image" (Ether, 3:15). It is the Father who is speaking. If, therefore, we can ascertain the form of the "Father of spirits," "The God of the spirits of all flesh," we shall be able to discover the form of the original man.

Jesus Christ, the Son of God, is "the express image" of His Father's person (Hebrews 1:3). He walked the earth as a human being, as a perfect man, and said, in answer to a question put to Him: "He that hath seen me hath seen the Father" (John 14:9). This alone ought to solve the problem to the satisfaction of every thoughtful, reverent mind. The conclusion is irresistible, that if the Son of God be the express image (that is, likeness) of His Father's person, then His Father is in the form of man; for that was the form of the Son of God, not only during His mortal life, but before His mortal birth, and after His resurrection. It was in this form that the Father and the Son, as two personages, appeared to Joseph Smith, when, as a boy of fourteen years, he received his first vision. Then if God made man--the first man--in His own image and likeness, he must have made him like unto Christ, and consequently like unto men of Christ's time and of the present day. That man was made in the image of Christ, is positively stated in the Book of Moses: "And I, God, said unto mine Only Begotten, which was with me from the beginning, Let us make man in our image, after our likeness; and it was so. * * * * And I, God, created man in mine own image, in the image of mine Only Begotten created I him, male and female created I them" (2:26, 27).

The Father of Jesus is our Father also. Jesus Himself taught this truth, when He instructed His disciples how to pray: "Our Father which art in heaven," etc. Jesus, however, is the firstborn among all the sons of God--the first begotten in the spirit, and the only begotten in the flesh. He is our elder brother, and we,

like Him, are in the image of God. All men and
women are in the similitude of the universal Father
and Mother, and are literally the sons and daughters
of Deity.

"God created man in His own image." This is
just as true of the spirit as it is of the body, which is
only the clothing of the spirit, its complement; the two
together constituting the soul. The spirit of man is in
the form of man, and the spirits of all creatures are in
the likeness of their bodies. This was plainly taught by
the Prophet Joseph Smith (Doctrine and Covenants,
77:2).

Here is further evidence of the fact. More than
seven hundred years before Moses was shown the
things pertaining to this earth, another great prophet,
known to us as the brother of Jared, was similarly
favored by the Lord. He was even permitted to behold
the spirit-body of the foreordained Savior, prior to His
incarnation; and so like the body of a man was His
spirit in form and appearance, that the prophet
thought he was gazing upon a being of flesh and
blood. He first saw the finger and then the entire body
of the Lord—all in the spirit. The Book of Mormon
says of this wonderful manifestation:

And it came to pass that when the brother
of Jared had said these words, behold, the Lord
stretched forth His hand and touched the
stones one by one with His finger; and the veil
was taken from off the eyes of the brother of
Jared, and he saw the finger of the Lord; and it
was as the finger of a man, like unto flesh and
blood; and the brother of Jared fell down before

the Lord, for he was struck with fear.

And the Lord saw that the brother of Jared had fallen to the earth; and the Lord said unto him, Arise, why hast thou fallen?

And he saith unto the Lord, I saw the finger of the Lord, and I feared lest he should smite me; for I knew not that the Lord had flesh and blood.

And the Lord said unto him, Because of thy faith thou hast seen that I shall take upon me flesh and blood; and never has man come before me with such exceeding faith as thou hast; for were it not so, ye could not have seen my finger. Sawest thou more than this?

And he answered, Nay, Lord, show thyself unto me.

And the Lord said unto him, Believest thou the words which I shall speak?

And he answered, Yea, Lord, I know that thou speakest the truth, for thou art a God of truth and canst not lie.

And when he had said these words, behold, the Lord showed himself unto him, and said, Because thou knowest these things ye are redeemed from the fall; therefore ye are brought back into my presence; therefore I show myself unto you.

Behold, I am He who was prepared from the foundation of the world to redeem my people. Behold, I am Jesus Christ, I am the Father and the Son. In me shall all mankind have light, and that eternally, even they who shall believe on my name; and they shall become my sons and my daughters.

And never have I shewed myself unto man whom I have created, for never hath man believed in me as thou hast. Seest thou that ye are created after mine own image? Yea, even all men were created in the beginning after mine own image.

Behold, this body, which ye now behold, is the body of my spirit, and man have I created after the body of my spirit; and even as I appear unto thee to be in the spirit, will I appear unto my people in the flesh. (Ether, 3:6-16.)

What more is needed to convince us that man, both in spirit and in body, is the image and likeness of God, and that God Himself is in the form of man?

When the divine Being whose spirit-body the brother of Jared beheld, took upon Him flesh and blood, He appeared as a man, having "body, parts and passions," like other men, though vastly superior to all others, because He was God, even the Son of God, the Word made flesh: in Him "dwelt the fulness of the Godhead bodily." And why should He not appear as a man? That was the form of His spirit, and it must needs have an appropriate covering, a suitable tabernacle. He came unto the world as He had promised to come (III Nephi, 1:13), taking an infant tabernacle, and developing it gradually to the fulness of His spirit stature. He came as man had been coming for ages, and as man has continued to come ever since. Jesus, however, as shown, was the only begotten of God in the flesh.

Adam, our great progenitor, "the first man," was, like Christ, a pre-existent spirit, and like Christ he took upon him an appropriate body, the body of a man, and so became a "living soul." The doctrine of the pre-existence, —revealed so plainly, particularly in latter days, pours a wonderful flood of light upon the otherwise mysterious problem of man's origin. It shows that man, as a spirit, was begotten and born of heavenly parents, and reared to maturity in the eternal mansions of the Father, prior to coming upon the earth in a temporal body to undergo an experience in mortality. It teaches that all men existed in the spirit before any man existed in the flesh, and that all who have inhabited the earth since Adam have taken bodies and become souls in like manner.

It is held by some that Adam was not the first man upon this earth, and that the original human being was a development from lower orders of the animal creation. These, however, are the theories of men. The word of the Lord declares that Adam was "the first man of all men" (Moses 1:34), and we are therefore in duty bound to regard him as the primal parent of our race. It was shown to the brother of Jared that all men were created in the *beginning* after the image of God; and whether we take this to mean the spirit or the body, or both, it commits us to the same conclusion: Man began life as a human being, in the likeness of our heavenly Father.

True it is that the body of man enters upon its career as a tiny germ or embryo, which becomes an infant, quickened at a certain stage by the spirit whose tabernacle it is, and the child, after being born, develops into a man. There is nothing in this, however, to indicate that the original man, the first of our race, began life as anything less than a man, or less than the human germ or embryo that becomes a man.

Man, by searching, cannot find out God. Never, unaided, will he discover the truth about the beginning of human life. The Lord must reveal Himself, or remain unrevealed; and the same is true of the facts relating to the origin of Adam's race—God alone can reveal them. Some of these facts, however, are already known, and what has been made known it is our duty to receive and retain.

The Church of Jesus Christ of Latter-day Saints, basing its belief on divine revelation, ancient and modern, proclaims man to be the direct and lineal off-spring of Deity. God Himself is an exalted man, per-fected, enthroned, and supreme. By His almighty power He organized the earth, and all that it contains, from spirit and element, which exist co-eternally with Himself. He formed every plant that grows, and every animal that breathes, each after its own kind, spiritu-ally and temporally—"that which is spiritual being in the likeness of that which is temporal, and that which is temporal in the likeness of that which is spiritual." He made the tadpole and the ape, the lion and the ele-phant; but He did not make them in His own image, nor endow them with Godlike reason and intelli-gence. Nevertheless, the whole animal creation will be perfected and perpetuated in the Hereafter, each class in its "distinct order or sphere," and will enjoy "eternal felicity." That fact has been made plain in this dispen-sation (Doctrine and Covenants, 77:3).

Man is the child of God, formed in the divine image and endowed with divine attributes, and even as the infant son of an earthly father and mother is capable in due time of becoming a man, so the undeveloped offspring of celestial parentage is capable, by experience through ages and aeons, of evolving into a God.

Joseph F. Smith,
John R. Winder,
Anthon H. Lund,

First Presidency of
The Church of Jesus Christ of Latter-day Saints.

From left to right: President Heber J. Grant, Second Counselor Charles W. Nibley, and First Counselor Anthony W. Ivins

Context

The 1909 First Presidency statement did not terminate discussions of evolution and human origins. The remainder of 1910 and the 1910-11 school year were years of considerable controversy at Brigham Young University, centering around so-called "higher criticism" (a method of scriptural interpretation, particularly of the Bible) and evolution. So in their annual Christmas message the First Presidency in December 1910 referred obliquely to the matter and counseled tolerance for divergent points of view. An excerpt from their message was thus included in the formal BYU Packet, introduced as follows:

In this Christmas message, the First Presidency devoted several sentences to the Church's position with regard to questions raised by science.

(2)
Words in Season
From the First Presidency
Deseret Evening News
Dec. 17, 1910, part 1, p. 3.

Diversity of opinion does not necessitate intolerance of spirit, nor should it embitter or set rational beings against each other. The Christ taught kindness, patience, and charity.

Our religion is not hostile to real science. That which is demonstrated, we accept with joy; but vain philosophy, human theory and mere speculations of men, we do not accept nor do we adopt anything contrary to divine revelation or to good common sense. But everything that tends to right conduct, that harmonizes with sound morality and increases faith in Deity, finds favor with us no matter where it may be found.

Context

Non-packet materials included hereafter (Statements of April 1910 and April 1911 are Appendix items A, B) provide additional insights about the intent of the preceding statements. But it was 1925 before the next document was produced which is included in the formal BYU Packet.

In 1925 occurred the famous Scopes trial in Dayton, Tennessee. John Thomas Scopes, a young high school science teacher, was being tried in court for his alleged breaking of Tennessee's law forbidding the teaching in the public schools of any version of creation "...that denies the story of the Divine Creation of man as taught in the Bible..." (Edward J. Larson, *Summer for the Gods: The Scopes Trial and America's Continuing Debate over Science and Religion* [New York: Basic Books, 1997], 50).

The trial was long known as "The World's Most Famous Court Case" and indeed that is the title of its published transcript. The trial ran from July 10 to July 21, 1925. In the midst thereof the LDS First Presidency released a formal statement. It was published July 18, 1925, in the *Deseret News* with the following explanation:

"Editor's Note: In connection with the wide-spread interest in evolution aroused by the famous trial in Tennessee, the First Presidency of the Church of Jesus Christ of Latter-day Saints, was recently invited by one of the foremost news organizations of the country, to outline the attitude of the Church on evolution. The following statement was prepared and appeared in many prominent newspapers throughout the United States over the signatures of Presidents Heber J. Grant, Anthony W. Ivins and Charles W. Nibley, First Presidency."

The statement was subsequently published in the September 1925 *Improvement Era* and from that source was included in the BYU Packet, as follows. It is noteworthy that the 1925 First Presidency chose not to send the November 1909 statement (Document 1) in response to the press inquiry. Instead they provided selected excerpts from that document. Readers will note that the shortened version produced in 1925 removes anti-science language from 1909. What was originally titled "The Origin of Man" in 1909 has now become, in shortened form, "'Mormon' View of Evolution."

(3)

Editors' Table: "Mormon" View of Evolution

Improvement Era
28 (September 1925): 1090-1091

A statement by the First Presidency of the Church of Jesus Christ of Latter-day Saints

"God created man in his own image, in the image of God created he him: male and female created he them."

In these plain and pointed words the inspired author of the book of Genesis made known to the world the truth concerning the origin of the human family. Moses, the prophet-historian, who "learned" we are told, "in all the wisdom of the Egyptians," when making this important announcement, was not voicing a mere opinion. He was speaking as the mouthpiece of God, and his solemn declaration was for all time and for all people. No subsequent revelator of the truth has contradicted the great leader and law-giver of Israel. All who have since spoken by divine authority upon this theme have confirmed his simple and sublime proclamation. Nor could it be otherwise. Truth has but one source, and all revelations from heaven are harmonious one with the other.

Jesus Christ, the Son of God, is "the express image" of his Father's person (Hebrews 1:3). He walked the earth as a human being, as a perfect man, and said, in answer to a question put to him: "He that hath seen me hath seen the Father" (John 14:9). This alone ought to solve the problem to the satisfaction of every thoughtful, reverent mind. It was in this form that the Father and the Son, as two distinct personages, appeared to Joseph Smith, when, as a boy of fourteen years, he received his first vision.

The Father of Jesus Christ is our Father also. Jesus himself taught this truth, when he instructed his disciples how to pray: "Our Father which art in heaven," etc. Jesus, however, is the first born among all the sons of God — the first begotten in the spirit, and the only begotten in the flesh. He is our elder brother, and we, like him, are in the image of God. All men and women are in the similitude of the universal Father and Mother, and are literally sons and daughters of Deity.

Adam, our great progenitor, "the first man," was, like Christ, a pre-existent spirit, and, like Christ, he took upon him an appropriate body, the body of a man, and so became a "living soul." The doctrine of pre-existence pours wonderful flood of light upon the otherwise mysterious problem of man's origin. It shows that man, as a spirit, was begotten and born of heavenly parents, and reared to maturity in the eternal mansions of the Father, prior to coming upon the earth in a temporal body to undergo an experience in mortality.

The Church of Jesus Christ of Latter-day Saints, basing its belief on divine revelation, ancient and modern, proclaims man to be the direct and lineal offspring of Deity. By his Almighty power God organized the earth, and all that it contains, from spirit and element, which exist co-eternally with himself.

Man is the child of God, formed in the divine image and endowed with divine attributes, and even as the infant son of an earthly father and mother is capable in due time of becoming a man, so the undeveloped offspring of celestial parentage is capable, by experience through ages and aeons, of evolving into a God.

Heber J. Grant,
Anthony W. Ivins,
Charles W. Nibley,

First Presidency.

Context

Non-packet materials transitioning from 1925 to the next Packet statement (1992) can be found in the Appendix, items C through F.

In the early 1990s a joint publishing agreement between the LDS Church and Macmillan Publishing Company resulted in the *Encyclopedia of Mormonism*. The entry on evolution is included in the BYU Packet since it had material input from the First Presidency. They authorized the inclusion of the excerpt from their minutes of 1931 and, of course, approved the inclusion of the entry itself in the BYU Packet. The 1931 minutes citation comes from the summary memo (April 7, 1931, see Appendix, item C) of the First Presidency to the other General Authorities, detailing a long-running discussion between certain brethren over two major doctrinal points. These concerned whether there had been death on this planet and human-like beings (so-called "pre-Adamites"), before the time usually ascribed to the Fall of Adam. The First Presidency ruled that, whatever the views of various brethren, the Church itself took no position either for or against either of these questions.

A brief summary of the provenance of the Encyclopedia article seems relevant. William Evenson, who as Associate Academic Vice-President of the university had helped arrange for BYU's role in the production of the *Encyclopedia of Mormonism*, was asked to write the article on Evolution due to his background as a scientist and his long interest in the subject. (He was one of several faculty members at BYU and elsewhere who had sufficient background to be considered for this task.) Evenson prepared several drafts of the article which became progressively longer under the guidance of Encyclopedia Associate Editor Noel B. Reynolds and with quite explicit feedback and direction from advisors to the Encyclopedia project, Apostles Neal A. Maxwell and Dallin H. Oaks.

A very few articles prepared for the Encyclopedia were taken to a meeting of the First Presidency and the Quorum of Twelve Apostles. When this article was considered, along with a significantly anti-evolution version prepared by another member of the Encyclopedia team, the Brethren determined, and First Counselor Gordon B. Hinckley directed, that only the brief excerpt of Evenson's article seen in the first half of the published Encyclopedia article, plus the excerpt from the 1931 First Presidency statement, should be included. It was made clear to Evenson that the Brethren did not want to imply either greater resolution on this issue than exists in the statements included in the *BYU Evolution Packet* or serious active disagreement among the Brethren themselves on this subject. Their position is still that which was expressed by the 1931 First Presidency, namely that this issue is not central to their calling and mission.

The Encyclopedia entry, the final document of the BYU Packet, is reproduced herewith.

(4)

"EVOLUTION"

WILLIAM E. EVENSON
Encyclopedia of Mormonism
(New York: Macmillan
Publishing, 1992), 2:478

The position of the Church on the origin of man was published by the First Presidency in 1909 and stated again by a different First Presidency in 1925:

The Church of Jesus Christ of Latter-day Saints, basing its belief on divine revelation, ancient and modern, declares man to be the direct and lineal offspring of Deity. . . . Man is the child of God, formed in the divine image and endowed with divine attributes (see Appendix, "Doctrinal Expositions of the First Presidency").

The scriptures tell why man was created, but they do not tell how, though the Lord has promised that he will tell that when he comes again (D&C 101:32-33). In 1931, when there was intense discussion on the issue of organic evolution, the First Presidency of the Church, then consisting of Presidents Heber J. Grant, Anthony W. Ivins, and Charles W. Nibley, addressed all of the General Authorities of the Church on the matter, and concluded,

> Upon the fundamental doctrines of the Church we are all agreed. Our mission is to bear the message of the restored gospel to the world. Leave geology, biology, archaeology, and anthropology, no one of which has to do with the salvation of the souls of mankind, to scientific research, while we magnify our calling in the realm of the Church. . . .
>
> Upon one thing we should all be able to agree, namely, that Presidents Joseph F. Smith, John R. Winder, and Anthon H. Lund were right when they said: "Adam is the primal parent of our race" [First Presidency Minutes, Apr. 7, 1931].

APPENDIX

OTHER AUTHORITATIVE
MATERIALS

Appendix Contents

Context

The method of the physical creation of the first humans has long been a matter of controversy among Latter-day Saint authorities and writers, with a wide range of views being expressed.

By 1895 President Wilford Woodruff found it necessary to direct Church members to avoid speculating on the issue (*Latter-Day Saints Millenial Star* *57:355*). But the November 1909 statement (cf. BYU Packet Document 1) apparently raised in the minds of many Church members the possibility that Adamic origins were again a permissible question. Queries on the topic to the First Presidency were answered five months later in the Presidency's official monthly column of instructions to priesthood quorums. Herein Church members are given three apparently acceptable options for the origin of the human body. Interestingly, the literal reading of scripture on this point is not included nor has any Church leader ever taught it. The entire statement is reproduced here.

(A)
Improvement Era
13 (April 1910): 570.

First Presidency

Priesthood Quorums' Table.

Origin of Man. — "In just what manner did the mortal bodies of Adam and Eve come into existence on this earth?" This question comes from several High Priests' quorums.

Of course, all are familiar with the statements in Genesis 1:26, 27; 2:7; also in the Book of Moses, Pearl of Great Price, 2:27; and in the Book of Abraham 5:7. The latter statement reads: "And the Gods formed man from the dust of the ground, and took his spirit (that is, the man's spirit) and put it into him; and breathed into his nostrils the breath of life, and man became a living soul."

These are the authentic statements of the scriptures, ancient and modern, and it is best to rest with these, until the Lord shall see fit to give more light on the subject. Whether the mortal bodies of man evolved in natural processes to present perfection, through the direction and power of God; whether the first parents of our generations, Adam and Eve, were transplanted from another sphere, with immortal tabernacles, which became corrupted through sin and the partaking of natural foods, in the process of time; whether they were born here in mortality, as other mortals have been, are questions not fully answered in the revealed word of God. For helpful discussion of the subject, see *Improvement Era*, Vol. XI, August 1908, No. 10, page 778, article, "Creation and Growth of Adam;" also article by the First Presidency, "Origin of Man," Vol. XIII, No. 1, page 75, 1909.

Context

As already indicated, 1910 and 1911 were contentious years at Brigham Young University, resulting in the non-retention of three faculty members. President Joseph F. Smith felt it necessary to explain his views on the matter to the Church at large. He published a signed editorial in the *Instructor*, the latter being the official magazine of the Church's Sunday School organization. It is notable for its expression of personal sentiments but also for its disavowal, in the last paragraph, of the existence of any official Church doctrine on the subject. It is included herewith in its entirety.

President Joseph F. Smith

(B)
Juvenile Instructor
46 (April 1911): 208-209.

President Joseph F. Smith

Philosophy and the Church Schools.

Some questions have arisen about the attitude of the Church on certain discussions of philosophy in the Church schools. Philosophical discussions, as we understand them, are open questions about which men of science are very greatly at variance. As a rule we do not think it advisable to dwell on questions that are in controversy, and especially questions of a certain character, in the courses of instruction given by our institutions. In the first place it is the mission of our institutions of learning to qualify our young people for the practical duties of life. It is much to be preferred that they emphasize the industrial and practical side of education. Students are very apt to draw the conclusion that whichever side of a controversial question they adopt is the truth, the whole truth, and nothing but the truth; and it is very doubtful, there-

fore, whether the great mass of our students have sufficient discriminating judgment to understand very
much about some of the advanced theories of philosophy or science.

Some subjects are in themselves, perhaps, per-
fectly harmless, and any amount of discussion over
them would not be injurious to the faith of our young
people. We are told, for example, that the theory of
gravitation is at best a hypothesis and that such is the
atomic theory. These theories help us to explain certain things about nature. Whether they are ultimately
true can not make much difference to the religious
convictions of our young people. On the other hand
there are speculations which touch the origin of life
and the relationship of God to his children. In a very
limited degree that relationship has been defined by
revelation, and until we receive more light upon the
subject we deem it best to refrain from the discussion
of certain philosophical theories which rather destroy
than build up the faith of our young people. One thing
about this so-called philosophy of religion that is very
undesirable, lies in the fact that as soon as we convert
our religion into a system of philosophy none but
philosophers can understand, appreciate, or enjoy it.
God, in his revelation to man, has made His word so
simple that the humblest of men, without especial
training, may enjoy great faith, comprehend the teachings of the Gospel, and enjoy undisturbed their reli-
gions convictions. For that reason we are averse to the
discussion of certain philosophical theories in our religious instructions. If our Church schools would con-
fine their so-called course of study in biology to that
knowledge of the insect world which would help us to

eradicate the pests that threaten the destruction of our crops and our fruit, such instruction would answer much better the aims of the Church school, than theories which deal with the origin of life.

These theories may have a fascination for our teachers and they may find interest in the study of them, but they are not properly within the scope of the purpose for which these schools were organized.

Some of our teachers are anxious to explain how much of the theory of evolution, in their judgment, is true, and what is false, but that only leaves their students in an unsettled frame of mind. They are not old enough and learned enough to discriminate, or put proper limitations upon a theory which we believe is more or less a fallacy. In reaching the conclusion that evolution would be best left out of discussions in our Church schools we are deciding a question of propriety and are not undertaking to say how much of evolution is true, or how much is false. We think that while it is a hypothesis, on both sides of which the most eminent scientific men of the world are arrayed, that it is folly to take up its discussion in our institutions of learning; and we can not see wherein such discussions are likely to promote the faith of our young people. On the other hand we have abundant evidence that many of those who have adopted in its fulness the theory of evolution have discarded the Bible, or at least refused to accept it as the inspired word of God. It is not, then, the question of the liberty of any teacher to entertain whatever views he may have upon this hypothesis of evolution, but rather the right of the Church to say that it does not think it profitable or wise to introduce controversies relative to evolution in its schools. Even if it were harmless from the standpoint of our faith, we think there are things more important to the daily affairs of life and the practical welfare of our young people. The Church itself has no philosophy about the *modus operandi* employed by the Lord in His creation of the world, and much of the talk therefore about the philosophy of Mormonism is altogether misleading. God has revealed to us a simple and effectual way of serving Him, and we should regret very much to see the simplicity of those revelations involved in all sorts of philosophical speculation. If we encouraged them it would not be long before we should have a theological scholastic aristocracy in the Church, and we should therefore not enjoy the brotherhood that now is, or should be common to rich and poor, learned and unlearned among the Saints.

JOSEPH F. SMITH.

Context

Despite President Smith's clear desires, the issue did not die. It surfaced again for the Scopes trial and First Presidency statement in 1925 (see BYU Packet Document 3), and then again about 1930.

Elder B. H. Roberts, the Church's leading defender and apologist and Senior President of the Seventy, had produced (under assignment) a study manual specifically for the Seventies of the Church but which was likely intended for the entire body of Melchizedek Priesthood holders. Titled *The Truth, The Way, The Life,* it attempted to synthesize truth, both scientific and religious, into one compact source. He concluded that death had been occurring on this planet for millions of years, and that human-like beings ("pre-Adamites") had lived on the planet for long periods before the opening of Adamic time. The formal reading committee was uncomfortable with those ideas, arguing that Roberts's stature was so great that including these in a Church manual would imply they were official doctrine. The committee thus felt it unwise to publish those views. Elder Joseph Fielding Smith, a junior member of the Quorum of the Twelve Apostles, vigorously promulgated an opposite point of view: that there had been no pre-Adamites and no death of any sort on the planet until after Adam's fall.

At length Elder Roberts was invited to present his views to the entire Quorum of the Twelve, which he did on January 7, 1931. Elder Smith requested and was permitted to make a rebuttal, which he did just two weeks later. The situation continued to simmer until the April 1931 General Conference, during which time a seven-page memo from the First Presidency to all the other General Authorities was produced. On April 7, 1931, the morning after General Conference, the First Presidency met with all the other Authorities and reviewed the Roberts/Smith dispute. They then ruled that the Church was not adopting a position on whether there was, or was not, death before the Fall of Adam, or whether there were, or were not, pre-Adamites. Neither side of either position was to be considered doctrine at all, and the Brethren were enjoined to leave the subjects totally alone.

The seven-page memo is what is referred to as "First Presidency Minutes" in the *Encyclopedia of Mormonism* entry on evolution (see BYU Packet Document 4), the excerpt from which was supplied by the 1992 First Presidency for inclusion in the article. We reproduce here the entire memo from the 1931 First Presidency (dated April 5, apparently the date on which it was typed for the April 7 meeting).

From left to right: First Counselor Anthony W. Ivins, President Heber J. Grant, Second Counselor Charles W. Nibley

(C)

Memo From the First Presidency
April 5, 1931

To the Council of the Twelve, the First Council of Seventy, and the Presiding Bishopric.

Dear Brethren:

On the 5th of April, 1930, at a conference of the Genealogical Society of Utah, Elder Joseph Fielding Smith delivered a sermon under the title "Faith Leads to a Fulness of Truth and Righteousness."

This sermon was published in the *Utah Genealogical and Historical Magazine*, and copies of it in pamphlet form were distributed, which gave it wide circulation.

In the sermon referred to, Elder Smith devotes the greater portion of his remarks to the subject of the creation of the earth and the relationship of our Father

Adam to it and its inhabitants. He refers to the conflict which exists between geologists and the scripture dates which are given, in regard to the period of time that has elapsed since the creation to the present, and definitely states that there was no death upon the earth, either vegetable, insect or animal, prior to the fall of man, and that human life did not exist upon the earth prior to Adam.

On the 15th of December, 1930, Elder B. H. Roberts submitted the following letter to the First Presidency:

"President Heber J. Grant, and Counselors; Building.

Dear Brethren:

I am writing you to ask if the article published in the *Utah Genealogical and Historical Magazine* of October, 1930, under the title "Faith Leads to a Fulness of Truth and Righteousness," dealing mainly with the antiquity of life and death upon the earth and treated as a discourse by Elder Joseph Fielding Smith on the 5th of April, 1930, is a treatise on that subject that was submitted to and approved by the Council of the First Presidency and perhaps the Quorum of the Twelve? And is it put forth as the official declaration of the Church on the subject treated? Or is it the unofficial and personal declaration of the opinion only of Elder Smith?

In the latter event then I feel that that fact should have been expressed in the discourse;

or if it is an official pronouncement of the Church then that fact should have been avowed; for the strictly dogmatical and the pronounced finality of the discourse demand the suggested explanation in either case.

If the discourse of Elder Smith is merely his personal opinion, while not questioning his right to such opinions, and also the right to express them, when avowed as his personal opinions, yet I object to the dogmatic and finality spirit of the pronouncement and the apparent official announcement of them, as if speaking with final authority.

If Elder Smith is merely putting forth his own opinions I call in question his competency to utter such dogmatism either as a scholar or as an Apostle. I am sure he is not competent to speak in such manner from general learning or special research work on the subject; nor as an Apostle, as in that case he would be in conflict with the plain implication at least of the scriptures, both ancient and modern, and with the teaching of a more experienced and learned and earlier Apostle than himself, and a contemporary of the Prophet Joseph Smith — whose public discourse on the subject appears in the *Journal of Discourses* and was publicly endorsed by President Brigham Young, all which would have more weight in setting forth doctrine than this last dictum of Elder Smith.

My question is important as affecting, finally, the faith and status of a very large portion of

the Priesthood and educated membership of
the Church, I am sure; and I trust the matter
will receive early consideration. All which is
respectfully submitted.

> Very truly your brother,
> (signed) B. H. Roberts"

The sermon referred to, with this letter, was
handed by the Presidency to the Council of Twelve
with the request that the matter be taken up, and the
difference of opinion which existed between the two
brethren be composed.

At a meeting of the Council of Twelve, Elder
Roberts was invited to be present and submit his find-
ings upon the question at issue, the principal point
involved being: Is the age of the earth greater than
that set forth in the scripture, as it is given in the Bible,
and was Adam the first human life upon it, or does he
represent the first of the human race that now occupy
it, and may human life have existed prior to his
advent.

Elder Roberts appeared before the Council of
Twelve and submitted a paper of fifty pages, in which
he quotes copiously from the sermon of Elder Smith,
and then proceeds to discuss the following statements
made in the sermon:

> "All life in the sea, on the earth, in the air,
> was without death. Things were not changing,
> as we find them changing in this mortal exis-
> tence, for mortality had not come. I denounce
> as absolutely false the opinion of some that this

earth was peopled by a race before Adam. I do not care what scientists say in regard to dinosaurs and other creatures upon the earth millions of years ago, that lived and died, and fought and struggled for existence."

Elder Roberts quotes from the scripture and extensively from the conclusions reached by the leading scientists of the world, to show that the earth is older than the time given to its creation in Genesis indicates. He places much stress upon the command of the Lord to Adam in which he says: "Be fruitful, and multiply, and replenish the earth." The word replenish he defines to mean to do a thing which has been done before, or refill that which has been made empty.

He quotes a statement made by Apostle Orson Hyde who, at a general conference of the Church, held October, 1854, declared that there were people upon the earth prior to the advent of Adam. Brigham Young and other of the presiding officers were present, and after the remarks made by Elder Hyde, President Young arose and said:

"I do not wish to eradicate any items from the lecture Elder Hyde has given us this evening, but simply to give you my views in a few words on the portion touching Bishops and Deacons. We have had a splendid address from Brother Hyde, for which I am grateful. I say to the congregation treasure up in your hearts what you have heard tonight, and at all other times."

Two weeks after Elder Roberts had submitted his paper Elder Smith appeared before the Council of Twelve and submitted a paper consisting of fifty-eight pages, in which he answers the arguments advanced by Elder Roberts, his contention being that Adam was the first man to come to this earth, and that consequently it could not have been previously inhabited by man; that there was no death upon the earth prior to the fall, neither vegetable, insect, or animal, which of course includes man.

In support of his argument he quotes extensively from the scripture, and from sermons of presiding men of the Church, particularly from the sermons of Orson Pratt, who refers to Adam as the first man, the first of all men, the Ancient of Days, etc. To meet the argument of Elder Roberts in the application of the word replenish he shows that the word may be used, and signifies, to fill as well as to fill again.

To meet the statement of Orson Hyde, Elder Smith says that Orson Hyde was not discussing the subject of Pre-Adamites, but was preaching upon marriage, and referred to Pre-Adamites incidentally. He admits that President Young was present, and that he endorsed the remarks made.

While there are many quotations cited by Elder Smith which refer to Adam as the first man, the following is the only one in which a pre-Adamic race is referred to. It is quoted under the heading: "Testimony of Charles W. Penrose":

"It is held by some that Adam was not the first man upon this earth, and that the original human being was a development from lower orders of the animal creation. These, however, are the theories of men. The word of the Lord declares that Adam was the first of all men, (Moses 1:34) and we are therefore in duty bound to regard him as the primal parent of our race.

                                        (signed)
                                    Joseph F. Smith
                                    John R. Winder
                                    Anthon H. Lund."

While this quotation is signed by the Presidency of the Church, it is given under the heading of "Testimony of President Charles W. Penrose."

After hearing granted to Elder Smith the following communication was received by the Presidency:

January 21, 1931

President Heber J. Grant & Counselors,
Dear Brethren:

We, the Council of the Twelve, to whom was referred the letter of Elder B. H. Roberts addressed to the First Presidency, a criticism of a certain discourse delivered by Elder Joseph Fielding Smith and published in the *Genealogical Magazine*, October, 1930, beg leave to report that we have given the time of three rather lengthy meetings to this matter.

At the first meeting Elder Roberts read and submitted a paper embodying his views at some length on the theory of pre-Adamic races, based on scientific investigation — a theory, we understand, which Elder Roberts has promulgated in some of his public utterances among the Latter-day Saints.

At the third meeting Elder Joseph Fielding Smith read and submitted a paper in which he defended the claim he made in the sermon published in the *Genealogical Magazine* above referred to, viz.; that pre-Adamic races on the earth is simply a theory and not a Church doctrine, and is not true. This he sought to prove by quoting Joseph Smith, the Prophet, Brigham Young, Parley P. Pratt, Orson Pratt, John Taylor and other high Church Authorities, particularly the late First Presidency, Joseph F. Smith, John R. Winder and Anthon H. Lund.

He also quoted a number of passages from the Bible, Book of Mormon, Doctrine & Covenants and Pearl of Great Price, pointing to the facts, as he construed them, that there were no pre-Adamic races of man on the earth, neither was there death upon the earth prior to the time of Adam.

We quote a sentence from Elder Roberts' letter: "If Elder Smith is merely putting forth his own opinions I call in question his competency to utter such dogmatism either as a scholar or as an Apostle. I am sure he is not competent to speak in such manner from general learning or special research work on the plain implication at least of the scriptures, both ancient and modern, and with the teaching of a more experienced and learned and earlier Apostle than himself, and a contemporary of the Prophet Joseph Smith."

This reference and language we regard as very offensive on the part of Elder Roberts, who fails to show the deference due from one brother to another brother of higher rank in the Priesthood. However, it may be said that these brethren affirmed at the close of the meeting that they entertained no ill feeling, one toward the other.

Elder Roberts' letter is herewith returned, and the two papers alluded to are now submitted to the Presidency. The Twelve await your further instructions relative to this matter, if you have any to give.

Sincerely your brethren,
The Council of the Twelve
By (signed) Rudger Clawson, Pres."

It will be observed that no suggestion is made in this communication regarding the attitude of the Council of Twelve in respect to the question involved in the controversy under consideration.

On February 9th the following communication was received from Elder Roberts:

"President Heber J. Grant and Counselors,
Building

Dear Brethren:

I feel almost as if I ought to apologize in
addressing this letter to you lest you think that
I am over-persistent in the representation of
things referred to herein.

You will recall that the letter I wrote to you
asking the questions in relation to the status of
Elder Joseph Fielding Smith's discourse published in the *Genealogical Magazine* for October
last, was referred to the Twelve for consideration. Agreeably to a request of theirs I submitted a paper (fifty typewritten pages) setting
forth precisely some of the objections I had to
the discourse. Two weeks later, bringing us to
January 21, Elder Smith submitted a paper of
about the same length to the Apostles, myself
being present. Since which time I have under-
stood that a report was made to the First
Presidency of which I have no copy. That is
now three weeks ago and just what the status
of the discussion or action upon it is I have not,
up to the present, learned.

The questions involved are of very great
importance from my standpoint. As for
instance, I would not like the matter to go to
judgment as matters now stand until I have an
opportunity to point out what to me are the
weakness and inconsistency of Elder Smith's
paper. There was really no discussion on the

subject before the Twelve, except the presentation of these two papers, and they represent solely the basis of discussion, not the discussion itself. And I have much more to present after hearing Elder Smith's reply to my paper, which should be said before any decision is rendered.

To me both the discourse on the points questioned and the paper in defense of them is slighter than a house of cards. Yet it was on such pabulum as this that suspended the publication of my book—now in manuscript— "The Truth, The Way, The Life!" This book from my judgment of it is the most important work that I have yet contributed to the Church, the six-volumed *Comprehensive History of the Church* not omitted.

Life at my years and with an incurable ailment is very precarious, and I should dislike very much to pass on without completing and publishing this work. I therefore ask that in any arrangement that may be made for a further hearing, I may be permitted to present my views on Elder Smith's paper in reply to mine, and if the position he has taken can be met successfully, then I think the principal cause of suspending the publication of my work, "The Truth, The Way, The Life" will be removed.

All which is respectfully submitted,

Very truly your brother,
(signed) B.H. Roberts"

After receipt of this latter communication the Presidency carefully reviewed the papers which had been submitted to the Council of Twelve, and after prayerful consideration decided that nothing would be gained by a continuation of the discussion of the subject under consideration.

The statement made by Elder Smith that the existence of pre-Adamites is not a doctrine of the Church is true. It is just as true that the statement "There were not pre-Adamites upon the earth", is not a doctrine of the Church. Neither side of the controversy has been accepted as a doctrine at all.

Both parties make the scripture and the statements of men who have been prominent in the affairs of the Church the basis of their contention; neither has produced definite proof in support of his views.

We quote the following from the *Millennial Star*, February 19, 1931:

"The sun is giving out energy daily. In a few million (or billion) years its energy will be gone. The other heavenly bodies are radiating and losing their heat; and in time they will be no better off than the age-bitten sun. The universe will run down. Then, on earth, there will be no summer and winter, perhaps no light and day, but just eternal twilight of middle African temperature, in the monotony of which all life will perish. So warns Sir James Jeans, famous British scientist, and brilliant writer and lecturer. Well for us that day is distant—a billion years or so—but, think of the grandchildren.

There is a ray of hope.

Dr. Robert A. Millikan, famous American scientist, and brilliant writer and lecturer, has discovered cosmic rays, sources of energy, that come from the uttermost confines of the universe to replenish the energy we lose by radiation. Out in the depths of space, by means unknown to us, the lost energy is assembled, converted, concentrated and sent back to delay the evil day. In short, Dr. Millikan says that this is a self-winding, self-repairing deathless universe. Day and night, summer and winter, may follow one another endlessly. That is more cheerful.

Whom are we to believe? These men are both world famous; both experimenters of the first rank, both honest men. Perhaps Dr. Millikan gives us a clue in his address as retiring president of the American Association for the Advancement of Science, delivered last Christmas week. He says:

'If Sir James Jeans prefers to hold one view and I another on this question, no one can say us nay. The one thing of which you may all be quite sure is that neither of us <u>knows</u> anything about it.'"

This is the frank and truthful admission of one of the foremost scientists of the world, an honest man, earnestly searching after truth, which he admits has not been definitely discovered.

The Prophet Joseph Smith said: "Oh, ye elders of Israel, hearken to my voice; and when you are sent into the world to preach, tell those things you are sent to tell; preach, and cry aloud,"Repent ye, for the kingdom of heaven is at hand; repent and believe the Gospel." Declare the first principles, and let mysteries alone, lest ye be overthrown. . . . Elder Brown, when you go to Palmyra say nothing about the four beasts, but preach those things the Lord has told you to preach about—repentance and baptism for the remission of sins."

We call attention to the fact that when one of the general authorities of the Church makes a definite statement in regard to any doctrine, particularly when the statement is made in a dogmatic declaration of finality, whether he express it as his opinion or not, he is regarded as voicing the Church, and his statements are accepted as the approved doctrines of the Church, which they should be.

Upon the fundamental doctrines of the Church we are all agreed. Our mission is to bear the message of the restored gospel to the people of the world. Leave Geology, Biology, Archaeology and Anthropology, no one of which has to do with the salvation of the souls of mankind, to scientific research, while we magnify our calling in the realm of the Church.

We can see no advantage to be gained by a continuation of the discussion to which reference is here made, but on the contrary are certain that it would lead to confusion, division and misunderstanding if carried further. Upon one thing we should all be able to agree, namely, that Presidents Joseph F. Smith, John R. Winder and Anthon H. Lund were right when they said: "Adam is the primal parent of our race."

Context

But there were still matters left unsettled. Elder James E. Talmage of the Twelve pointed out that Joseph Fielding Smith's position had been made clear in a speech widely published the previous year (*Utah Genealogical and Historical Magazine* 21 (October 1930): 145-158); that of Elder Roberts had never been made public. It could thus be construed, wrongly, that Elder Smith's views were the official Church position. Apparently the Brethren concurred. So in August 1931 Elder Talmage delivered a speech in the Tabernacle which asserted forcefully that death had indeed occurred on this planet for millions of years before the time of Adam, and that there had indeed been pre-Adamites (though he did not think they were ancestral to modern humans). And the speech, as was customary, was sent to the *Church News* for publication.

Elder Smith immediately insisted that the Talmage speech should not be published, and the First Presidency took his sentiments under consideration. The question (of publication) was considered variously over the next three months, at which time the First Presidency ruled that the speech should indeed be published in the *Church News*. It appeared there

James E. Talmage in his laboratory.

November 17, 1931. But the First Presidency took a further step: the Talmage speech was to be published also as a separate pamphlet *over the Church's imprimatur*, i.e., published *by* the Church. For that reason we include it in this collection, though readers should recognize that Talmage's scientific comments reflect data of 1931 and are not an accurate summary for the present day.

(D)

THE EARTH AND MAN

Address Delivered in the Tabernacle,
Salt Lake City, Utah
Sunday, August 9, 1931, By
DR. JAMES E. TALMAGE
of the Council of the Twelve Apostles

Published by the Church of Jesus Christ
of Latter-day Saints

Reprint from The Deseret News, Salt Lake City,
Utah, November 21, 1931
Printed in the United States of America

*"In the beginning God created the heaven and the earth.
"And the earth was without form, and void; and darkness
was upon the face of the deep. And the Spirit of God moved
upon the face of the waters." (Gen. 1:1, 2.)*

Any question as to when that beginning was is
largely futile because unanswerable. In the first place
we have no time unit by which to measure back

through the ages to the time at which, so far as the earth is concerned, time began.

Years are as inadequate in any attempted survey of the stages of earth development as are miles to the astronomer who would span the distances of inter-stellar space. He speaks in terms of light-years, such unit being the distance traversed by a ray of light speeding on at the rate of approximately 180,000 miles per second throughout a year.

Secondly, we are without information as to what stage of earth development is indicated by "the begin-ning." And what is a beginning in nature? At best it is but a new start in advance of what had passed up to that point of time; and every beginning is an ending of what went immediately before, even as every con-summation is a commencement of something greater, higher, and therefore superior to the past.

THE EARTH OLDER THAN MAN

To the thoughtful mind there can be no confu-sion of the beginning spoken of in the opening verse of Genesis with the advent of man upon the changing earth; for by the scriptural record itself we learn of stage after stage, age after age of earth processes by which eventually this planet became capable of sup-porting life—vegetable, animal and human in due course.

Whether or not scientists have been able to see, however dimly, the way by which the earth as an orb in space was formed, matters little except as a subject of academic interest. For many years it was very gen-erally believed that the earth, once formless and void,

passed through stages of cooling of superheated gas
to liquid, thence to the solid state, as the Nebular
Theory assumed; but this conception has given way to
the later thought that the earth as a solid spheroid has
resulted from the bringing together of particles once
diffused in space—this being the basis of the
Planetesimal Hypothesis.

But this we know, for both revealed and discovered truth, that is to say both scripture and science,
so affirm—that plant life antedated animal existence
and that animals preceded man as tenants of earth.

LIFE AND DEATH BEFORE MAN'S ADVENT

According to the conception of geologists the
earth passed through ages of preparation, to us
unmeasured and immeasurable, during which count-
less generations of plants and animals existed in great
variety and profusion and gave in part the very substance of their bodies to help form certain strata which
are still existent as such.

The oldest, that is to say the earliest, rocks thus
far identified in land masses reveal the fossilized
remains of once living organisms, plant and animal.
The coal strata, upon which the world of industry so
largely depends, are essentially but highly com-
pressed and chemically changed vegetable substance.
The whole series of chalk deposits and many of our
deep-sea limestones contain the skeletal remains of
animals. These lived and died, age after age, while the
earth was yet unfit for human habitation.

FROM THE SIMPLE TO THE COMPLEX

From the fossil remains of plants and animals found in the rocks the scientist points to a very definite order in the sequence of life embodiment, for the older rocks, the earlier formations, reveal to us organisms of simplest structure only, whether of plants or animals. These primitive species were aquatic; land forms were of later development. Some of these simpler forms of life have persisted until the present time, though with great variation as the result of changing environment.

Geologists say that these very simple forms of plant and animal bodies were succeeded by others more complicated; and in the indestructible record of the rocks they read the story of advancing life from the simple to the more complex, from the single-celled protozoan to the highest animals, from the marine algae to the advanced types of flowering plant—to the apple-tree, the rose, and oak.

What a fascinating story is inscribed upon the stony pages of the earth's crust! The geologist, who through long and patient effort has learned at least a little of the language in which these truths are written, finds the pages illustrated with pictures, which for fidelity of detail excel the best efforts of our modern engravers, lithographers and half-tone artists. The pictures in the rocks are the originals, the rest at best but copies.

In due course came the crowning work of this creative sequence, the advent of man! Concerning this all-important event we are told that scientists and theologians are at hopeless and irreconcilable variance. I regard the assumption or claim, whichever it be, as an exaggeration. Discrepancies that trouble us now will diminish as our knowledge of pertinent facts is extended. The Creator has made record in the rocks for man to decipher; but He has also spoken directly regarding the main stages of progress by which the earth has been brought to be what it is. The accounts can not be fundamentally opposed; one can not contradict the other; though man's interpretation of either may be seriously at fault.

ADAM A HISTORIC PERSONAGE

So far as the history of man on the earth is concerned the scriptures begin with the account of Adam. True, the geologist does not know Adam by name; but he knows and speaks of man as an early, continuing and present form of earth-life, above and beyond all other living things past or present.

We believe that Adam was a real personage, who stands at the head of his race chronologically. To my mind Adam is a historic personage, not a prehistoric being, unidentified and uncertain.

If the Usher [sic] chronology be correct, or even approximately so, then the beginning of Adamic history as recorded in scripture dates back about 4000 years before the birth of Christ. We as a Church believe that the current reckoning of time from the birth of Christ to the present is correct, namely 1931 years—not from last New Year's day, January 1, but from the month that came to be known among the Hebrews as Nisan or Ahib, corresponding with our late March and early April. So we believe that we are now living in the 1931st year since the birth of Christ, and therefore 5931 years since the beginning of the Adamic record.

This record of Adam and his posterity is the only scriptural account we have of the appearance of man upon the earth. But we have also a vast and ever-increasing volume of knowledge concerning man, his early habits and customs, his industries and works of art, his tools and implements, about which such scriptures as we have thus far received are entirely silent. Let us not try to wrest the scriptures in an attempt to explain away what we can not explain. The opening chapters of Genesis, and scriptures related thereto, were never intended as a text-book of geology, archeology, earth-science or man-science. Holy Scripture will endure, while the conceptions of men change with new discoveries. We do not show reverence for the scriptures when we misapply them through faulty interpretation.

PRIMARY AND SECONDARY CAUSES

There has been much discussion over the alleged conflict between the teachings of science and the doctrines of the revealed word concerning the origin of man. Let it be remembered that the term origin is almost invariably used in a relative sense. The mind of man is unable to grasp the fundamental thought of an absolute or primary origin. Every occurrence man has witnessed is the result of some previously acting cause or purpose; and that cause in turn was the effect or result of causes yet more remote. Perhaps we have never been able to trace an effect to its primary or original cause. Man may say that he understands the origin of an oak in the acorn from which it sprang; but is not the acorn the fruit of a yet earlier oak, and so in reality rather a continuation than a beginning? Yet there is something fascinating in the thought of a beginning; the persistence of a process once started is far less mysterious than its inception.

It is not enough to refer effects to the *First Great Cause;* it is unsatisfying and not always reverent to answer questions as to how things came to be what they are by the easy statement that God made them so. With such an answer the scientific man has little patience. The fact that all created things are the works of God and that all processes of nature are due to Him as the administrator of law and order is to the scientific mind an axiom requiring neither argument nor demonstration. The botanist knows that God makes the plant grow: but he, weak mortal, is devoting time and the energy of body, mind and spirit, to a study of the way in which God works such a marvelous miracle. The geologist knows that God created the earth; but the best effort of his life is put forth in the hope of finding out in some degree, however small, the method by which the Creator wrought this wondrous world. The astronomer gazing into the starry depths sees in their orderly procession the Lord Eternal walking in His majesty and might; and in humility the student of the heavenly bodies spends days and nights striving to learn a little of the way in which God worked out the marvel of the universe.

In proportion as any one of these may learn of the ways of God he becomes wise. To be able to think as God thinks, to comprehend in any degree His purposes and methods, is to become in that measure like unto Him, and to that extent to be prepared for eventual companionship in His presence. The scientist is busily engaged in the study of secondary causes—the ways and means by which God works and through which He accomplishes His miracle, ever beginning, never ending—and in his search for the truth the student of science scarcely dares lift his eyes to look toward the First Great Cause, the Eternal Power that stands and operates behind and above all the secondary causes, or what we call the processes of Nature.

THE ORIGIN OF MAN

The question involved in the origin of man therefore, is not raised as a challenge to the belief and declaration that he came to earth through Divine direction, but is in the nature of an inquiry as to the conditions under which he came. There are many who claim that man's advent upon the earth was effected through processes of evolution from lower forms, processes that had been operative for ages, processes by which man is made kin to the brute and a development from the lowest type of organism. Others affirm that he differs from all mortal creatures of lower rank, not only in degree but in kind; in short, that he is not one with the animal creation and that therefore his coming was in no sense a natural and necessary result of earlier animal life. Discussion on this question has developed intense animus, and too often the quest for truth has been lost sight of in the strife for triumph.

In speaking of the origin of man, we generally have reference to the creation of man's body; and, of all the mistakes that man has made concerning himself, one of the greatest and the gravest is that of mistaking the body for the man. The body is no more truly the whole man than is the coat the body. The man, as an individual intelligence, existed before his earthly body was framed and shall exist after that body has suffered dissolution. Let it not be assumed that belief in the existence of man's spirit is a conception founded upon scriptural authority only; on the contrary, let it be known that it is in accordance with the best and most advanced scientific thought and philosophic belief of the day to hold that man consists of spirit and body; and Divine revelation makes plain that these together constitute the soul.

We have difficulty in comprehending processes for which we find no analogy in things familiar. Even were it possible for us to know in detail the way in which the body of man was formed and then endowed with the power of procreation, insuring the perpetuity of the race, it would throw but little light upon the subject of the ultimate origin of man. We know but little of things beyond the sphere upon which we live except as information has been revealed by a power superior to that of earth, and by an intelligence above that of man. Notwithstanding the assumption that man is the culmination of an evolutionary development from a lower order of beings, we know that the body of man today is in the very form and fashion of his spirit, except indeed for disfigurements and deformities. The perfect body is the counterpart of the perfect spirit and the two are the constituent entities of the soul.

BY WHAT STANDARD?

Much depends upon the standard by which we judge as to whether any particular organism shall be pronounced of high or lower rank. By the standard of powers of flight, in which the bird excels, man is a very inferior being; if judged by fleetness of foot he is far below the deer; by gage of strength he is inferior to the horse and the elephant; and yet man holds dominion over these and all other living things of earth. In certain important points of body-structure man stands low in the scale if he be graded strictly in accordance with the accepted standard of mammalian anatomy.

In the course of creative events the earth came to a condition fitted for the abiding place of the sons and daughters of God; and then Adam came forth upon the earth. But the beginning of man's mortal existence upon the earth was not the beginning of man; he had lived before, even as he shall live after the earth has passed away and its place taken by a new earth and a new heaven.

MAN AND THE APE

It has been stated by certain extremists that evolution affirms that man is in the line of posterity from the ape. But scientists today discredit this view. The most that even radical evolutionists assert is that the similarity of structure between man and certain apes indicates the possibility of a common ancestor of the two; but between man and the ape there are more essential differences than resemblances.

True, man does not excel in strength of limb, agility, or speed, but in the God-given powers of mind and in the possession of superior ambition and effort. Hear the words of one who until his death was regarded as among the foremost of American geologists, James D. Dana:

> *"Man's origin has thus far no sufficient explanation from science.* His close relations in structure to the man-apes are unquestionable. They have the same number of bones with two exceptions, and the bones are the same in kind and structure. The muscles are mostly the same. Both carry their young in their arms. The affiliations strongly suggest community of descent. But the divergencies . . . especially the cases of degeneracy in man's structure, exhibited in his palmigrade feet and the primitive character of his teeth, allying him in these respects to the Lower Eocene forms, are admitted proof that he has not descended from any existing type of ape. In addition, man's erect posture makes the gap a very broad one. The brute, the ape included, has powerful muscles in the back of the neck to carry the head in its horizontal position, while man has no such muscles, as any one of the species can prove by crawling for a while on 'all fours.' Beyond this, the great size of the brain, his eminent intellectual and moral qualities, his voice and speech, give him sole title to the position at the head of the kingdoms of life. In this high position, he is able to use Nature as his work-mate, his com-

panion, and his educator, and to find perpetual delight in her harmonies and her revelations.
. . .

"Whatever the results of further search,
we may feel assured, in accord with Wallace,
who shares with Darwin in the authorship of
the theory of Natural Selection, that *the intervention of a Power above Nature was at the basis of
man's development.* Believing that Nature exists
through the will and ever-acting power of the
Divine Being, and that all its great truths, its
beauties, its harmonies, are manifestations of
His wisdom and power, or, in the words near-
ly of Wallace, that the whole universe is not
merely dependent on, but actually is, the will
of one Supreme Intelligence, Nature, with man
as its culminant species, is no longer a mys-
tery."

James D. Dana, *Manual of Geology*, 4th edition,
page 1036.

These lines were written shortly before the
death of the writer — and constitute his last testament
and testimony as to the origin of the species to which
he himself belonged.

MAN'S PLACE IN NATURE

In the work already cited. the same author
wrote:

"Man stands in the successional line of the

quadrumana, at the head of the animal kingdom. But he is not a primate among primates.
The quadrumana are, as Cuvier called them,
quadrumana from the first to the last. They are
brute mammals, as is manifested in their carni-
vore-like canines and their powerful jaws; in
their powerful muscular development; in their
walking on all fours, and the adaptation thereto exhibited in the vertebrae, producing the
convexity of the back; and also in other parts of
the skeleton. Man, on the contrary, is not
quadrumanous. . . .

"Man was the first being, in the geological
succession, capable of an intelligent survey of
Nature and a comprehension of her laws; the
first capable of augmenting his strength
by bending Nature to his service, rendering
thereby a weak body stronger than all possible
animal force; the first capable of deriving hap-
piness from truth and goodness; of appre-
hending eternal right; of reaching toward a
knowledge of self and God; the first, therefore,
capable of conscious obedience or disobedi-
ence of a moral law, and the first subject to
debasement of his moral nature through his
appetites.

"There is in man, therefore, a spiritual element in which the brute has no share. His
power of indefinite progress, his thoughts and
desires that look onward even beyond time,
his recognition of spiritual existence and of a
Divinity above, all evince a nature that par-
takes of the infinite and divine. Man is linked

to the *past* through the system of *life*, of which he is the last, the completing, creation. But, unlike other species of that closing system of the *past*, he, through his *spiritual* nature, is more intimately connected with the opening *future*." —

Dana, pages 1017-18.

A LATER AUTHORITY

Let me cite a later authority than Dana. Among the living no anthropologist has been more pronounced in upholding the theories of Darwin and Lamarck than Dr. Henry Fairfield Osborn.

By the theories mentioned man was said to have risen from tree-climbing ape-like ancestors. In his address as retiring president of the American Association for the Advancement of Science, December, 1929, Dr. Osborn affirms the untenability of the views he had so long and aggressively advocated. He regards the human bones unearthed at Piltdown, Sussex, England, as typical of the "Dawn Man," who was in every distinguishing characteristic, a *man*, not part man and part ape, but as to brain capacity and other evidences of mentality equal to some races now living. Yet Osborn holds to a communal origin of man and anthropoids related in structure, away back in the late Tertiary age of geologic history.

Thus theories come, endure for a season and go, like the fungi of the night; nevertheless they serve their purpose as temporary aids in human thought and endeavor.

THE TIME ELEMENT

The outstanding point of difference between those who take the opening chapters of Genesis and cognate scriptures as the whole and only reliable record of the creation of earth and man, and the students of earth-science who fail to find an adequate record in scripture, is the point of time during which man in some state has lived on this planet.

Geologists and anthropologists say that if the beginning of Adamic history dates back but 6000 years or less, there must have been races of human sort upon earth long before that time — without denying, however, that Adamic history may be correct, if it be regarded solely as the history of the Adamic race.

This view postulates, by application of Dana's affirmation already quoted: "*that the intervention of a power above Nature*" brought about the placing of, let me say, Adam upon earth.

It is but fair to say that no reconciliation of these opposing conceptions has been effected to the satisfaction of both parties. We have not yet learned how to correlate geologic time-periods with terms of years, except as estimates, for which no absolutely dependable foundation may be found.

NOBILITY OF ADAM'S RACE

I do not regard Adam as related to—certainly not as descended from—the Neanderthal, the Cro-Magnon, the Peking or the Piltdown man. Adam came as divinely directed, created and empowered, and stands as the patriarchal head of his posterity—a posterity, who, if true to the laws of God, are heirs to the Priesthood and to the glories of eternal lives.

Were it true that man is a product of evolution from lower forms, it is but reasonable to believe that he will yet develop into something higher. While it is a fact that eternal progression is a characteristic of man's Divine birthright, as yet we have learned nothing to indicate that man shall develop physically into any other form than that in which he now appears.

Many attempts have been made by those who regard man as an animal to frame some definition by which he may be distinctively described among his fellow animals; but of such attempts none have been satisfactorily successful. The difficulty lies in the fact, already stated, that man differs from the animal creation not only in degree but in kind; he is the only being who has any conception of a preexistent state or an existence beyond the grave; the only being whose thoughts turn toward God and who feels in his soul the inspiring impulses of kinship to Deity. Believe not those who would make man but little above the brutes, when in truth he is but little below the angels, and if faithful shall pass by the angels and take his place among the exalted sons of God. The spirit of man is the offspring of the Eternal Father, and his body, if unmarred, is in the very form and fashion of that spirit.

THE ANTE-MORTAL STATE

We have been told that Jesus Christ is in very truth our Elder Brother, and as to His preexistence in the spirit state there is little room for question. That His spirit was in the form of the earthly body which He afterward took and which body was slain, buried, and resurrected, and with which body He ascended, into heaven, is attested by scripture. Going back to the time immediately following the dispersion from Babel, we read of a prophet to whom the unembodied Lord revealed Himself, saying: "Behold, this body, which ye now behold, is the body of my spirit; and man have I created after the body of my spirit; and even as I appear unto thee to be in the spirit will I appear unto my people in the flesh." (*Book of Mormon,* Ether, 3:16.)

It is evident from this scripture that in His pre-existent state, that is to say in the state in which He existed prior to His earthly birth, Jesus Christ had the same form and stature that He afterward presented in the flesh. By natural processes His spirit shaped for itself a body from the material of earth, which body underwent a course of graded development until it reached maturity, in which state that body was the counterpart of the spirit whose material tabernacle it was. As with Jesus, so with all the sons and daughters of God; each had a spiritual existence before he entered upon this stage of mortal existence, and in each case the body is formed and fashioned by the power of the immortal spirit. In this process of body-shaping, the spirit may be hindered, hampered, and interfered with, through influences of heredity, through prenatal defects, or through accident and disease.

As to how were formed the bodies of the first human beings to take tabernacles, the revealed word gives no details while science has practically nothing to offer by way of explanation. As Dana so positively declares in the work already cited, *"Man's origin has thus far no sufficient explanation from science."*

Man's mortal existence is but temporary to this earth; he came hither from another realm, in which he lived in an unembodied state and to which, in the natural order, he shall return in disembodied state, following the change known as death. After the body of the first man had been made ready through the direct operation of the creative power, the spirit of man entered that body. Note the sublimity of the scriptural declaration; "And the Lord God formed man of the dust of the ground, and breathed into his nostrils the breath of life; and man became a living soul." (Gen. 2: 7.)

A POWER ABOVE NATURE

In the study of all the created things over which he has dominion, man has found it possible to investigate with some degree of success the secondary causes, or natural processes through which the creative power has operated to bring about the system that we designate as Nature; but in the study of his own eternal self he is brought at once to the contemplation of the First Great Cause as to his origin. The power that lies at the basis of man's development is "a Power above Nature." That is to say, man, as a mortal being, exists as the result of a special and particular creation. Through graded stages the earth was brought into a state suited to the support of life. In orderly sequence plants and animals appeared; and when at last the world was prepared for its royal ruler, he came, even as had been declared.

"And God said, Let us make man in our image, after our likeness; and let them have dominion over the fish of the sea, and over the fowl of the air, and over the cattle, and over all the earth, and over every creeping thing that creepeth upon the earth.

"So God created man in his own image, in the image of God created he him; male and female created he them.

"And God blessed them, and God said unto them, Be fruitful, and multiply, and replenish the earth, and subdue it: and have dominion over the fish of the sea, and over the fowl of the air, and over every living thing that moveth upon the earth." (Gen. 1: 26-28.)

Such is the declaration of scripture regarding Adam's advent upon earth; and such is a fair summary of our knowledge upon the subject.

EVOLUTION, TRUE AND FALSE

Evolution is true so far as it means development, and progress, and advancement in all the works of God; but many of the vagaries that have been made to do duty under that name are so vague as to be unacceptable to the scientific mind. At best, the conception of the development of man's body from the lower forms through evolutionary processes has been but a theory, an unproved hypothesis. Theories may be regarded as the scaffolding upon which the builder stands while placing the blocks of truth in position. It is a grave error to mistake the scaffolding for the wall, the flimsy and temporary structure for the stable and permanent. The scaffolding serves but a passing purpose, important though it be, and is removed as soon as the walls of that part of the edifice of knowledge have been constructed. Theories have their purpose and are indispensable, but they must never be mistaken for demonstrated facts. The Holy Scriptures should not be discredited by theories of men; they can not be discredited by fact and truth. Within the Gospel of Jesus Christ there is room and place for every truth thus far learned by man, or yet to be made known. The Gospel is not behind the times; on the contrary it is up-to-date and ever shall be.

It is natural for the young and immature mind to think that what to it is new must of necessity be new to the world. Comparatively inexperienced students are discovering from time to time apparent discrepancies between the faith of their fathers and the development of modern thought; and these they are apt to magnify and exaggerate, when as a matter of fact, their great-grandfathers met the same seeming difficulties and yet survived. Believe not those who assert that the Gospel of Jesus Christ is in any way opposed to progress or inconsistent with advancement.

IN THE LINEAGE OF DEITY

Man is the child of God, he is born heir to boundless possibilities, the inheritor of the eternities to come. Among mortal beings, the law holds true that the posterity of each shall be after his kind. The child therefore may become like unto the parent; and man may yet attain the rank of godship. He is born in the lineage of Deity, not in the posterity of the brute creation.

I cite my words of an earlier day, with a quotation.

MAN'S RELATIVE LITTLENESS

The insignificance of man in comparison with the earth on which he dwells, and even with limited topographical features of his world, has ofttimes been dwelt upon. Draw to scale a towering mountain and a man standing at its base or on its summit — what does the man amount to? But then the earth as a planet is small compared with some others of its own system, to say nothing of the relative sizes of earth and sun. In turn, our entire solar system, in the measurement of which miles cease to have meanings — so vast it is — ranks low in dimensions as we gage it with other families of worlds in the great galaxy of stars to which it belongs, and that immeasurable galaxy is but one among many, and not the greatest of them all.

DREAM VISION OF THE INFINITE

This hour is not well suited to the presentation of mathematical data relating to the extent of the universe; though it may permit us to indulge the contemplation of thought-pictures, bewildering though that indulgence may be. John Paul Richter's *Dream Vision of the Infinite* has been brought to English readers through several renditions; and I ask you to follow or accompany me through one of these, generally worded along the lines of the version given us by Thomas De Quincey:

"God called up from dreams a man into the vestibule of heaven, saying 'Come thou hither and I will show thee the glories of my house.' And to the servants that stood around the throne He said 'Take the man and strip from him his robes of flesh; cleanse his vision and put a new breath into his nostrils; only touch not with any change his human heart — the heart that fears and trembles.'

"It was done; and, with a mighty angel for his guide, the man stood ready for his infinite voyage. Then, from the terraces of heaven, without sound or farewell, they wheeled away into endless space. Sometimes, with solemn flight of angel wing, they fled through Zaarrahs of darkness, through wildernesses of death that divided the worlds of life. Sometimes they swept over frontiers that were quickening under prophetic motions from God.

"Then, from a distance that is counted only in heaven, light dawned for a time through a sleepy film. By unutterable pace the light swept to them, they by unutterable pace to the light. In a moment the rushing of planets was upon them; in a moment the blazing of suns was around them.

"Then came eternities of twilight, that revealed, but were not revealed. To the right hand and the left towered mighty constellations, that by self-repetitions and answers from afar, that by counterpositions, built up triumphal gates, whose architraves, whose archways — horizontal, upright — rested, rose — at altitudes, by spans — that seemed ghostly from infinitude. Without measure were the architraves, past number were the archways, beyond memory the gates!

"Within were stairs that scaled the eternities above, that descended to the eternities below: above was below, below was above, to the man stripped of gravitating body. Depth was swallowed up in height insurmountable; height was swallowed up in depth unfathomable. Suddenly, as thus they rode from infinite to infinite, suddenly as thus they tilted over abysmal worlds, a mighty cry arose — that systems more mysterious, that worlds more billowy, other heights and other depths were coming, were nearing, were at hand!

"Then the man sighed and stopped, shuddered and wept. His overladen heart uttered itself in tears; and he said 'Angel, I will go no farther; for the spirit of man aches with this infinity. Insufferable is the glory of God. Let me lie down in the grave and hide myself from the persecutions of the infinite; for end, I see, there is none!'

"And from all the listening stars that shone around issued a choral chant, 'The man speaks truly; end is there none that ever yet we heard of.' 'End is there none?' the angel solemnly demanded. 'Is there, indeed, no end? And is this the sorrow that kills you?' Then the angel threw up his glorious hands to the heaven of heavens, saying 'End is there none to the universe of God! Lo, also, there is no beginning!'"

THE SPIRITUAL GRANDEUR OF MAN

What is man in this boundless setting of sublime splendor? I answer you: Potentially now, actually to be, he is greater and grander, more precious according to the arithmetic of God, than all the planets and suns of space. For him were they created; they are the handiwork of God; man is His son! In this world man is given dominion over a few things; it is his privilege to achieve supremacy over many things.

"The heavens declare the glory of God; and the firmament showeth His handiwork." (Psa. 19:1.) Incomprehensibly grand as are the physical creations of the earth and space, they have been brought into existence as means to an end, necessary to the realization of the supreme purpose, which in the words of the Creator is thus declared:

"For behold, this is my work and my glory — to bring to pass the immortality and eternal life of man." (Pearl of Great Price, page 4.)

It is decreed that this earth shall become a celestialized, glorified sphere; such is the revealed word. Science has nothing to say on the matter; it can neither refute nor prove. But the Lord, even God, hath spoken it — and so shall it be! Amen.

Context

The Talmage speech was also published in the Church's official magazine for the Sunday Schools, the *Instructor*, 100, No. 12 (December 1965): 474-477, and 101, No. 1 (January 1966): 9-11,15. The order of topics was considerably rearranged in this publication, however, without editorial notification to the reader. Roberts's manuscript *The Truth, The Way, The Life* was eventually published by two different publishers, with a series of analytical essays by modern scholars, in 1994. The essay by James B. Allen titled "The Story of *The Truth, The Way, The Life*" gives an excellent summary of the manuscript's history. See *The Truth, The Way, The Life: An Elementary Treatise on Theology*, 2d ed., edited by John W. Welch, (Provo, Utah: BYU Studies, 1996).

Interested readers will find further related information in the book *Can Science Be Faith-Promoting?* edited by Stanley Larson (Salt Lake City, UT: Blue Ribbon Books, 2001). This work contains some sixty-seven pages of correspondence between Elder James E. Talmage's eldest son and intellectual/spiritual heir, Sterling B. Talmage, and various Church leaders—primarily Elder Talmage, Elder

Joseph Fielding Smith, Elder John A. Widtsoe, and President Heber J. Grant.

There has been some minor manufactured controversy over whether this speech (Document D) was really approved by the First Presidency for publication. They were not adopting it as doctrine, of course, only ensuring that a position of neutrality was maintained. In his 1931 journal, Elder Talmage summarized their decision as follows.

(E)

Elder James E. Talmage

Journal,
November 16, 17, 1931, 29:67

"Nov. 16, Mon. I was called into brief consultation by the First Presidency on the subject of my Tabernacle address of August 9."

"Nov. 17, Tues. According to appointment made yesterday the First Presidency gave special attention to the matter of my Tabernacle address before referred to, going over it with considerable care, though it was apparent to me that the brethren had before considered it among themselves and had reached their decision. This they announced to me by way of instruction to send back the copy which I had recalled from the printer, and to have the address published in the *Deseret News* of next Saturday evening, and further to have it printed in pamphlet form. I shall make further comment when the address is actually in print."

Context

His "further comment," a detailed summary of the history summarized above, appears in his journal the day the address was published, November 21, 1931.

And to alleviate any concern as to Elder Talmage's objectivity, we also include here the summary of the President of the Church on the matter.

(F)

President Heber J. Grant

Journal,
November 16, 17, 1931

"Monday, Nov. 16, 1931. Brother James E. Talmage called and the Presidency went over his sermon delivered in the Tabernacle two or three months ago, which had been referred to the Apostles. They had not been able to agree on it. We asked him to go over his sermon and make some slight changes that had been suggested by the brethren, which he agreed to do. The matter has been in abeyance as to its publication for many weeks."

"17 Nov. 1931. At 11:30 Brother James E. Talmage called, and we went over his address delivered in the Tabernacle a number of weeks ago, and authorized its publication and also gave authorization for it to be printed in the same form as the radio addresses, for distribution."

(pp. 218, 219)

Context

Church Presidents and First Presidencies fell silent on the issue for many years, though a few unofficial publications appeared on occasion. President David O. McKay, however, touched on the subject briefly in a speech delivered at Brigham Young University in 1952.

(G)

President David O. McKay

"There is a perpetual design permeating all purposes of creation. On these thoughts, science again leads the student up to a certain point and sometimes leads [leaves] him with his soul unanchored. Milikan (sic) is right when he says, "Science without religion obviously may become a curse rather than a blessing to mankind." But, science dominated by the spirit of religion is the key to progress and the hope of the future. For example, evolution's beautiful theory of the creation of the world offers many perplexing problems to the inquiring mind. Inevitably, a teacher who denies divine agency in creation, who insists there is no intelligent purpose in it, will infest [infect] the student with the thought that all may be chance. I say, that no youth should be so led without a counter-balancing thought. Even the skeptic teacher should be fair enough to see that even Charles Darwin, when he faced this great question of annihilation, that the creation is dominated only by chance wrote: "It is an intolerable thought that man and all other sentient beings are doomed to complete annihilation after such long, continued slow progress." And another good authority, Raymond West, said, "Why this spiniture [expenditure] of time and pain and blood?" Why should man come so far if he's destined to go no farther? A creature that travels such distances and fought such battles and won such victories deserves what we are compelled to say, "To conquer death and rob the grave of its victory." The public school teacher will probably, even if he says that much, will go no farther. In the Church school the teacher is unhampered. In the Brigham Young University and every other church school the teacher can say God is at the helm."

President David O. McKay
"A Message for L.D.S. College Youth"
speech at Brigham Young University
Oct. 8, 1952
BYU Speeches of the Year, pp. 5-6

Note: An edited version of the above passage was included in a General Conference address of Pres. McKay, read by his son David Lawrence McKay. Among other alterations, the word 'beautiful' is deleted in that version. *Conference Reports*, April 1968, p. 92.

President David O. McKay

Context

In 1954 a major anti-evolution book was published which some readers interpreted as an authoritative Church statement of doctrine. President McKay responded to numerous inquiries from Church educators and members that the book contained the author's views only and was not approved by the Church. (See Gregory A. Prince and Wm. Robert Wright, *David O. McKay and the Rise of Modern Mormonism* [Salt Lake City: University of Utah Press, 2005] 45-49). But eventually, without denouncing the work and its author publicly, President David O. McKay was moved to make a brief statement at Brigham Young University.

(H)

President David O. McKay

"Whatever the subject may be, the principles of the gospel of Jesus Christ may be elaborated upon without fear of anyone's objecting, and the teacher [at BYU] can be free to express his honest conviction regarding it, whether that subject be in geology, the history of the world, the millions of years that it took to prepare the physical world, whether it be in engineering, literature, art—any principles of the gospel may be briefly or extensively touched upon for the anchoring of the student who is seeking to know the truth."

President David O. McKay
"Gospel Ideals—Life's Surest Anchor"
Speech at BYU
Oct. 30, 1956
BYU Speeches of the Year, p. 6

Context

No substantive further comments by either a Church President or First Presidency are known to us until the remarks of President Spencer W. Kimball in the October 1975 Women's Conference, when he spoke of Adam and Eve.

(I)

President Spencer W. Kimball

"And I, God created man in mine own image, in the image of mine Only Begotten created I him; male and female created I them. [The story of the rib, of course, is figurative.]. . . ."

"Man became a living soul – mankind, male and female. The Creators breathed into their nostrils the breath of life and man and woman became living souls. We don't know exactly how their coming into this world happened, and when we're able to understand it the Lord will tell us."

President Spencer W. Kimball
"The Blessings and Responsibilities of
Womanhood"
Ensign 6:71, 72
March 1976
(Bracketed material in original)

Context

No further authoritative statements appeared until the *Encyclopedia of Mormonism* was published in 1992 (see BYU Packet Document 4). President Gordon B. Hinckley became President of the Church in 1995, and in 1997 he delivered a speech to students at the LDS Institute in Ogden, Utah. The speech is a series of responses to students' questions, exhorting them to good living and high commitments, building to a response about evolution, and then to personal testimony. A variety of interpretations can be sustained in the context of the overall speech; we suggest readers consult the speech in its entirety. The kernel statement about evolution is given here.

(J)

President Gordon B. Hinckley

"People ask me every now and again if I believe in evolution. I tell them I am not concerned with organic evolution. I do not worry about it. I passed through that argument long ago."

Discourses of President Gordon B. Hinckley
Deseret Book Co.,
Salt Lake City, Utah
Speech given April 15, 1997
commercial edition v. 1 p. 379
commemorative edition v. 1 p. 463

Context

About this time President Hinckley granted an interview to a journalist, Larry A. Witham, who was preparing a book on science and religion in America. The writer inquired about Mormonism's views toward evolution, and reported President Hinckley's statement, which he assured us was accurately transcribed from his tape-recorded interview.

(K)

President Gordon B. Hinckley

"What the church requires is only belief 'that Adam was the first man of what we would call the human race,' says Gordon Hinckley, the church's living prophet. Scientists can speculate on the rest, he says, recalling his own study of anthropology and geology: 'Studied all about it. Didn't worry me then. Doesn't worry me now.'"

Quoted in Larry A. Witham,
Where Darwin Meets the Bible:
Creationists and Evolutionists in America
(New York: Oxford University Press, 2002),
p. 177.

Context

A slightly expanded and more detached comment from Elder Hinckley appears in the May 2004 *New Era* (p. 37). But since it comes from a speech given more than a decade before he became President of the Church, we have not included it here.

Persons inquiring of the First Presidency during the years after 1992 have, as has been Church policy, traditionally been given answer by a letter to their ecclesiastical leaders, which generally included a shortened version of the *Encyclopedia of Mormonism* article (see BYU Packet Document 4). In 2003 and early 2004 such inquirers were referred to the Church Public Relations department which in turn furnished (on occasion, not consistently) by one means or another the following statement.

(L)

Responses to Inquiries to First Presidency, ca. 2003-2004

" 'God created man in his own image, in the image of God created he him; male and female created he them.' In these plain and pointed words the inspired author of the book of Genesis made known to the world the truth concerning the origin of the human family."

Readers will recognize these sentences as the opening to the third paragraph of BYU Packet Document 1 (November 1909) and the first paragraph of BYU Packet Document 3 (September 1925).

Context

The sentiments expressed in these "Responses to Inquiries" do indeed seem to summarize the critical aspects of the issue. The method of creation is left unresolved.

Afterword

As indicated above, the documents included in this publication were selected according to their authoritative status for expressing the official position of the LDS Church on biological evolution. Every document of which we are aware that meets the criteria for authoritative status is included, irrespective of the position expressed on the subject of evolution. Three categories of authoritative status were identified for inclusion here:

1. Formal statements by the First Presidency as a body: Documents 1, 2, 3, and Appendix item C.

According to the 1992 BYU Board of Trustees, consisting of the First Presidency and seven of the Twelve Apostles, such formal statements are the definitive source of official Church positions. The BYU Packet was limited to statements in this category plus the *Encyclopedia of Mormonism* article on evolution, as specifically authorized by the First Presidency in 1992.

2. Documents authorized for distribution by the First Presidency: Cover Letter of BYU Packet, Document 4, and Appendix items A, D, and L. Also included are

supporting documents E and F from the Appendix.

These include the *Encyclopedia of Mormonism* article on evolution; the talk, "The Earth and Man," by Elder James E. Talmage and items E and F relating to its publication by the Church as authorized by the First Presidency; an April 1910 instruction to Priesthood Quorums, and 2003-2004 responses to inquiries to the First Presidency.

3.   Documents after 1909 from a President of the Church addressing the subject directly in his capacity as President: Appendix items B, G, H, I, J, and K.

The first formal First Presidency statement (Category 1 above) was published in 1909 and provided an authoritative foundation for all subsequent statements. For this reason, we took 1909 as the starting point for our compilation of statements by Presidents of the Church.

We know of two other documents which quite likely meet the above criteria.   These are "The Relatedness of Living Things" by Bertrand F. Harrison, *The Instructor* 100, No. 7 (July 1965): 272-276; and "An Official Position" by William Lee Stokes, *Dialogue* 12, No. 4 (Fall 1979): 90-92. Extensive documentation exists that these were approved for publication by President David O. McKay; see, for example, Gregory A. Prince and Wm. Robert Wright, *David O. McKay and the Rise of Modern Mormonism*, (Salt Lake City: University of Utah Press, 2005), 49. But these are not included here because we do not have ultimate documentation from President McKay's records directly.

A number of private letters from Church Presidents or First Presidencies have seen limited circulation but are not included here since they were never openly published or approved for publication by the principals involved.

Numerous other publications in LDS literature address these issues. Despite the widespread distribution and popularity of some, they remain as expressions of their authors' views only. They do not constitute statements of official doctrine, and in some cases they incurred specific private criticism from the Church President or First Presidency. None of them has reflected the full range of authoritative expressions on this topic.

It is hoped that the present publication will be useful in putting the authoritative statements into a more readily available context and format.

Index

William Evenson is an administrator and physics professor at Utah Valley State College, where he moved in 2004 upon retiring from BYU. During his thirty-four years at BYU Evenson was Professor of Physics, Dean of General Education, Dean of Physical and Mathematical Sciences, and Associate Academic Vice President. He and his wife, Nancy, live in Provo, and keep in touch with their five children and a growing number of grandchildren spread across the United States.

Duane E. Jeffery is Professor of Integrative Biology at Brigham Young University. He has published professionally in various biological journals and on matters of Mormonism and science. He has received numerous teaching awards at BYU including Honors Professor of the Year and the Karl G. Maeser Distinguished Teaching Award. He serves as a member of the Board of Directors of the National Center for Science Education. He and his wife, Kaye, live in American Fork, Utah.